THE TRAINING OF THE HUMAN PLANT

THE TRAINING OF THE HUMAN PLANT

BY

LUTHER BURBANK

GYAN

Published by

GYAN

Gyan Publishing House
5, Ansari Road
Daryaganj, New Delhi-110002
Phone: 011-47034999, 9811692060
E-mail: books@gyanbooks.com

Distribution Network
gyanbooks.com
India, USA, Canada, UK, Australia

ISBN: 978-93-6208-378-4 (HB)
First Published, 1907

2nd Impression 2024

Printed at: Gyan Press, Delhi.

THE TRAINING OF THE HUMAN PLANT
Author: LUTHER BURBANK

THE TRAINING OF THE HUMAN PLANT

THE TRAINING OF THE HUMAN PLANT

BY

LUTHER BURBANK

NEW YORK
THE CENTURY CO.
1922

DEDICATED

TO THE

TWENTY MILLION

PUBLIC SCHOOL CHILDREN OF AMERICA

AND TO THE

UNTOLD MILLIONS

UNDER OTHER SKIES

CONTENTS

THE TRAINING OF THE HUMAN PLANT

THE TRAINING OF THE HUMAN PLANT

I

THE MINGLING OF RACES

DURING the course of many years of investigation into the plant life of the world, creating new forms, modifying old ones, adapting others to new conditions, and blending still others, I have constantly been impressed with the similarity between the organization and development of plant and human life. While I have never lost sight of the principle of the survival of the fittest and all that it implies as an explanation of the development and progress of plant life, I have come to find in the crossing of species and in selection, wisely directed, a great and powerful instrument for the transformation of the vegetable kingdom along lines that lead constantly upward. The crossing of species is to me paramount. Upon it, wisely directed and accompanied by a rigid selection of the best and as rigid an exclusion of the poorest, rests the hope of all progress. The mere crossing of species, unaccompanied by selection, wise supervision, intelligent care, and the utmost patience, is not likely to result in marked good, and may result in vast harm. Unorganized effort is often most vicious in its tendencies.

Before passing to the consideration of the adaptation of the principles of plant culture and improvement in a more or less modified form to the human being, let me lay emphasis on the opportunity now presented in the United States for observing and, if we are wise, aiding in what I think it fair to say is the grandest opportunity ever presented of developing the finest race the world has ever known out of the vast mingling of races brought here by immigration.

By statistical abstract on immigration, prepared by the Bureau of Statistics of the Department of Commerce and Labor in Washington, I find, that, in the year 1904, 752,864 immigrants came into the United States, assigned to more than fifty distinct nationalities. It will be worth while to look carefully at this list. It shows how widely separated geographically, as well as ethnologically, is the material from which we are drawing in this colossal example of the crossing of species:

Austria-Hungary, including Bohemia, Hungary, and other Austria save Poland	117,156
Belgium	3,976
Denmark	8,525
France	9,406
Germany	46,380
Greece	11,343
Italy	193,296
Netherlands	4,916
Norway	23,808
Poland	6,715
Rumania	7,087
Russia	145,141
Spain	3,996
Sweden	27,763
Switzerland	5,023
Carried forward . . .	614,531

Brought forward . .	614,531	
* Turkey in Europe . .	5,669	
England	38,620	
Ireland	36,142	
Scotland	11,092	
Wales	1,730	
Europe not specified . .	143	
Total Europe . .		707,927
British North America .	2,837	
Mexico	1,009	
Central America . . .	714	
West Indies and Miguelon	10,193	
South America	1,667	
Total America . . .		16,420
China	4,309	
Japan	14,264	
Other Asia	7,613	
Total Asia . . .		26,186
Total Oceania . .		1,555
Total Africa . . .		686
All other countries .		90
Total Immigrants .		752,864

* Includes Servia, Bulgaria, and Montenegro.

Study this list from any point of view. Where has there been found a broader opportunity for the working out of these underlying principles? Some of these immigrants will mate with others of their own class, notably the Jews, thus not markedly changing the current; many will unite with others of allied speech; still others marry into races wholly different from their own, while a far smaller number will perhaps find union with what we may call native stock.

But wait until two decades have passed, until there are children of age to wed, and then see, under the changed conditions, how widespread will be the mingling. So from the first the foreign nations have been pouring into this country and taking their part in this vast blending.

Now, just as the plant breeder always notices sudden changes and breaks, as well as many minor modifications, when he joins two or more plants of diverse type from widely separated quarters of the globe,—sometimes merging an absolutely wild strain with one that, long over-civilized, has largely lost virility,—and just as he finds among the descendants a plant which is likely to be stronger and better than either ancestor, so may we notice constant changes and breaks and modifications going on about us in this vast combination of races, and so may we hope for a far stronger and better race if right principles are followed, a magnificent race, far superior to any preceding it. Look at the material on which to draw! Here is the North, powerful, virile, aggressive, blended with the luxurious, ease-loving, more impetuous South. Again you have the merging of a cold phlegmatic temperament with one mercurial and volatile. Still again the union of great native mental strength, developed or undeveloped, with bodily vigor, but with inferior mind. See, too, what a vast number of environmental influences have been at work in social relations, in climate, in physical surroundings. Along with this we must observe the merging of the vicious with the good, the good with the good, the vicious with the vicious.

II

THE TEACHINGS OF NATURE

WE are more crossed than any other nation in the history of the world, and here we meet the same results that are always seen in a much-crossed race of plants: all the worst as well as all the best qualities of each are brought out in their fullest intensities. Right here is where selective environment counts. When all the necessary crossing has been done, then comes the work of elimination, the work of refining, until we shall get an ultimate product that should be the finest race ever known. The best characteristics of the many peoples that make up this nation will show in the composite: the finished product will be the race of the future.

In my work with plants and flowers I introduce color here, shape there, size or perfume, according to the product desired. In such processes the teachings of nature are followed. Its great forces only are employed. All that has been done for plants and flowers by crossing, nature has already accomplished for the American people. By the crossings of types, strength has in one instance been secured; in another, intellectuality; in still another, moral force. Nature alone has done this. The work of man's head and hands has not yet been summoned to prescribe for the development of a race. So far a preconceived and mapped-out crossing of bloods finds no place in the making of peoples and nations. But when nature has already done its duty, and the crossing leaves a product which in the rough displays the best human attributes, all that is left to be done falls to selective environment.

But when two different plants have been crossed, that is only the beginning. It is only one step, however important; the great work lies beyond—the care, the nurture, the influence of surroundings, selection, the separation of the best from the poorest, all of which are embraced in the words I have used—selective environment.

How, then, shall the principles of plant culture have any bearing upon the development of the descendants of this mighty mingling of races?

All animal life is sensitive to environment, but of all living things the child is the most sensitive. Surroundings act upon it as the outside world acts upon the plate of the camera. Every possible influence will leave its impress upon the child, and the traits which it inherited will be overcome to a certain extent, in many cases being even more apparent than heredity. The child is like a cut diamond, its many facets receiving sharp, clear impressions not possible to a pebble, with this difference, however, that the change wrought in the child from the influences without becomes constitutional and ingrained. A child absorbs environment. It is the most susceptible thing in the world to influence, and if that force be applied rightly and constantly when the child is in its most receptive condition, the effect will be pronounced, immediate, and permanent.

Where shall we begin? Just where we begin with the plant, at the very beginning. It has been said that the way to reform a man is to begin with his grandfather. But this is only a half-truth; begin with his grandfather, but begin with the grandfather when he is a child. I find the following quoted from the great kindergartner Froebel:

> "The task of education is to assist natural development toward its destined end.
>
> "As the beginning gives a bias to the whole after development, so the early beginnings of education are of most importance."

While recognizing the good that has been accomplished in the early kindergarten training of children, I must enter a most earnest protest against beginning education, as we commonly use the word, at the kindergarten age. No boy or girl should see the inside of a school-house until at least ten years old. I am speaking now of the boy or girl who can be reared in the only place that is truly fit to bring up a boy or a plant —the country, the small town or the country, the nearer to nature the better. In the case of children born in the city and compelled to live there, the temptations are so great, the life so artificial, the atmosphere so like that of the hothouse, that the child must be placed in school earlier as a matter of safeguarding.

But, some one asks, How can you ever expect a boy to graduate from college or university if his education does not begin until he is ten years of age? He will be far too old.

First I answer that the curse of modern child-life in America is over-education. For the first ten years of this, the most sensitive and delicate, the most pliable life in the world, I would prepare it. The properly prepared child will make such progress that the difference in time of graduation is not likely to be noticeable; but, even if it should be a year or two later, what real difference would it make? Do we expect a normal plant to begin bearing fruit a few weeks after it is born? It must have time, ample time, to be prepared for the work before it. Above all else, the child must be a healthy animal. I do not work with diseased plants. They do not cure themselves of disease. They only spread disease among their fellows and die before their time.

III

DIFFERENTIATION IN TRAINING

I WISH to lay special stress upon the absurdity, not to call it by a harsher term, of running children through the same mill in a lot, with absolutely no real reference to their individuality. No two children are alike. You cannot expect them to develop alike. They are different in temperament, in tastes, in disposition, in capabilities, and yet we take them in this precious early age, when they ought to be living a life of preparation near to the heart of nature, and we stuff them, cram them, and overwork them until their poor little brains are crowded up to and beyond the danger-line. The work of breaking down the nervous systems of the children of the United States is now well under way. It is only when some one breaks absolutely away from all precedent and rule and carves out a new place in the world that any substantial progress is ever made, and seldom is this done by one whose individuality has been stifled in the schools. So it is imperative that we consider individuality in children in their training precisely as we do in cultivating plants. Some children, for example, are absolutely unfit by nature and temperament for carrying on certain studies. Take certain young girls, for example, bright in many ways, but unfitted by nature and bent, at this early age at least, for the study of arithmetic. Very early,—before the age of ten, in fact,—they are packed into a room along with from thirty to fifty others and compelled to study a branch which, at best, they should not undertake until they have reached maturer years. Can one by any possible cultivation and selection and crossing compel figs to grow on thistles or apples on a banana-tree? I have made many varied and strange plant combinations in the hope of betterment and still am at work upon others, but one cannot hope to do the impossible.

THE FIRST TEN YEARS

NOT only would I have the child reared for the first ten years of its life in the open, in close touch with nature, a barefoot boy with all that implies for physical stamina, but should have him reared in love. But you say, How can you expect all children to be reared in love? By working with vast patience upon the great body of the people, this great mingling of races, to teach such of them as do not love their children to love them, to surround them with all the influences of love. This will not be universally accomplished to-day or to-morrow, and it may need centuries; but if we are ever to advance and to have this higher race, now is the time to begin the work, this very day. It is the part of every human being who comprehends the importance of this to bend all his energies toward the same end. Love must be at the basis of all our work for the race; not gush, not mere sentimentality, but abiding love, that which outlasts death. A man who hates plants, or is neglectful of them, or who has other interests beyond them, could no more be a successful plant-cultivator than he could turn back the tides of the ocean with his finger-tips. The thing is utterly impossible. You can never bring up a child to its best estate without love.

BE HONEST WITH THE CHILD

THEN, again, in the successful cultivation of plants there must be absolute honesty. I mean this in no fanciful way, but in the most practical and matter-of-fact fashion. You cannot attempt to deceive nature or thwart her or be dishonest with her in any particular without her knowing it, without the consequences coming back upon your own head. Be honest with your child. Do not give him a colt for his very own, and then, when it is a three-year-old, sell it and pocket the proceeds. It does not provoke a tendency in children to follow the Golden Rule, and seldom enhances their admiration and respect for you. It is not sound business policy or fair treatment; it is not honest. Bear in mind that this child-life in these first ten years is the most sensitive thing in the world; never lose sight of that. Children respond to ten thousand subtle influences which would leave no more impression upon a plant than they would upon the sphinx. Vastly more sensitive is it than the most sensitive plant. Think of being dishonest with it!

Hére let me say that the wave of public dishonesty which seems to be sweeping up over this country is chiefly due to a lack of proper training—breeding, if you will—in the formative years of life. Be dishonest with a child, whether it is your child or some other person's child—dishonest in word or look or deed, and you have started a grafter. Grafting, or stealing,—for that is the better word,—will never be taken up by a man whose formative years have been spent in an atmosphere of absolute honesty. Nor can you be dishonest with your child in thought. The child reads your motives as no other human being reads them. He sees into your own heart. The child is the purest, truest thing in the world. It is absolute truth: that 's why we love children. They know instinctively whether you are true or dishonest with them in thought as well as in deed; you cannot escape it. The child may not always show its knowledge, but its judgment of you is unerring. Its life is stainless, open to receive all impressions, just as is the life of the plant, only far more pliant and responsive to influences, and to influences to which no plant is capable of being responsive. Upon the child before the age of ten we have an unparalleled opportunity to work; for nowhere else is there material so plastic.

TRAITS IN PLANTS AND BOYS

TEACH the child self-respect; train it in self-respect, just as you train a plant into better ways. No self-respecting man was ever a grafter. Make the boy understand what money means, too, what its value and importance. Do not deal it out to him lavishly, but teach him to account for it. Instil better things into him, just as a plant-breeder puts better characteristics into a plant. Above all, bear in mind repetition, repetition, the use of an influence over and over again. Keeping everlastingly at it, this is what fixes traits in plants—the constant repetition of an influence until at last it is irrevocably fixed and will not change. You cannot afford to get discouraged. You are dealing with something far more precious than any plant —the priceless soul of a child.

KEEP OUT FEAR

AND, again, keep fear out that the child may grow up to the end of the first ten-year period and not learn what physical fear is. Let him alone for that, if he is a healthy normal child; he will find it and profit by it. But keep out all fear of the brutal things men have taught children about the future. I believe emphatically in religion. It has been said "God made religion, and man made theology," just as "God made the country, and man made the town." I have the largest sympathy for religion, and the largest contempt I am capable of for a misleading theology. Do not feed children on maudlin sentimentalism or dogmatic religion; give them nature. Let their souls drink in all that is pure and sweet. Rear them, if possible, amid pleasant surroundings. If they come into the world with souls groping in darkness, let them see and feel the light. Do not terrify them in early life with the fear of an after-world. Never was a child made more noble and good by the fear of a hell. Let nature teach them the lessons of good and proper living, combined with an abundance of well-balanced nourishment. Those children will grow to be the best men and women. Put the best in them by contact with the best outside. They will absorb it as a plant absorbs the sunshine and the dew.

IV

SUNSHINE, GOOD AIR AND NOURISHING FOOD

WE cannot carry a great plant-breeding test to a successful culmination at the end of a long period of years without three things, among many others, that are absolutely essential—sunshine, good air, and nourishing food.

SUNSHINE

TAKE the first, both in its literal and figurative sense—sunshine. Surround the children with every possible cheer.

I do not mean to pamper them, to make them weak; they need the winds, just as the plants do, to strengthen them and to make them self-reliant. If you want your child to grow up into a sane, normal man, a good citizen, a support of the state you must keep him in the sunshine. Keep him happy. You cannot do this if you have a sour face yourself. Smiles and laughter cost nothing. Costly clothing, too fine to stand the wear and tear of a tramp in the woods or sliding down a haystack or a cellar door, are a dead weight upon your child. I believe in good clothes, good strong serviceable clothes for young children—clothes that fit and look well; for they tend to mental strength, to self-respect. But there are thousands of parents who, not having studied the tremendous problems of environmental surroundings, and having no conception of the influence of these surroundings, fail to recognize the fact that either an over-dressed or a poorly dressed child is handicapped.

Do not be cross with the child; you cannot afford it. If you are cultivating a plant, developing it into something finer and nobler, you must love it, not hate it; be gentle with it, not abusive; be firm, never harsh. I give the plants upon which I am at work in a test, whether a single one or a hundred thousand, the best possible environment. So should it be with a child, if you want to develop it in right ways. Let the children have music, let them have pictures, let them have laughter, let them have a good time; not an idle time, but one full of cheerful occupation. Surround them with all the beautiful things you can. Plants should be given sun and air and the blue sky; give them to your boys and girls. I do not mean for a day or a month, but for all the years. We cannot treat a plant tenderly one day and harshly the next; they cannot stand it. Remember that you are training not only for to-day, but for all the future, for all posterity.

FRESH AIR

To develop indoors, under glass, a race of men and women of the type that I believe is coming out of all this marvelous mingling of races in the United States is immeasurably absurd. There must be sunlight, but even more is needed, fresh, pure air. The injury wrought to-day to the race by keeping too young children indoors at school is beyond the power of any one to estimate. The air they breathe even under the best sanitary regulations is far too impure for their lungs. Often it is positively poisonous—a slow poison which never makes itself fully manifest until the child is a wreck. Keep the child outdoors and away from books and study. Much you can teach him, much he will teach himself all gently, without knowing it, of nature and nature's God, just as the child is taught to walk or run or play; but education in the academic sense shun as you would the plague. And the atmosphere must be pure around it in the other sense. It must be free from every kind of indelicacy or coarseness. The most dangerous man in the community is the one who would pollute the stream of a child's life. Whoever was responsible for the saying that "boys will be boys" and a young man "must sow his wild oats" was perhaps guilty of a crime.

NOURISHING FOOD

It is impossible to apply successfully the principles of cultivation and selection of plants to human life if the human life does not, like the plant life, have proper nourishment. First of all, the child's digestion must be made sound by sufficient, simple, well-balanced food. But, you say, any one should know this. True, and most people do realize it in a certain sense; but how many realize that upon the food the child is fed in these first ten years largely depends its moral future? I once lived near a class of people who, from religious belief, excluded all meat, eggs, and milk from the dietary of their children. They fed them vegetables and the product of cereals. What result followed? The children were anemic, unable to withstand disease, quickly succumbed to illness. There were no signs of vigor; they were always low in vitality. But that was not all. They were frightfully depraved. They were not properly fed; their ration was unbalanced.[1] Nature re-

[1] The request has often come to me to state what I thought a "well-balanced" food especially for children. We all need food which supplies the elements of *growth* and *repair* and all, both old and young, must also have foods which yield *warmth* and *energy*. Nearly all foods contain both these elements though in greatly

belled; for she had not sufficient material to perfect her higher development.

What we want in developing a new plant, making it better in all ways than any of its kind that have preceded it, is

varying proportions and usually far from the right ones for growth and health unless a variety of foods are eaten at each meal. Growing children need a greater proportion of body-building foods, such as lean meats, fish, milk, some vegetables and fruits. They are often fed *too great a proportion of sweet and starchy foods.* A certain proportion of these are absolutely necessary but we all know the "starch babies" by their pale, fat, flabby, characterless faces, lusterless eyes and general lack of vitality. Less starchy foods and some fresh meats with eggs, milk, vegetables and plenty of fruits will give more vitality, a better growth, greater intelligence, better health and a better constitution, notwithstanding the belief of some of my vegetarian friends to the contrary.

Children mostly fed on sweet and farinaceous foods are also starved for the various *salts* and *mineral elements.* These must all be supplied especially to children else they will certainly become victims of an unbalanced, unnatural, premature development and a shortening of life simply from starvation. Life, the builder, must have the necessary materials or the structure must be imperfect and incomplete.

L. B.

a splendid norm, not anything abnormal. So we feed it from the soil, and it feeds from the air by the aid of sunlight and thus we make it a powerful aid to man. It is dependent upon good food. Upon good food for the child, well-balanced food, depends good digestion; upon good digestion, with pure air to keep the blood pure, depends the nervous system. If you have the first ten years of a boy's or a girl's life in which to make them strong and sturdy with normal nerves, splendid digestion, and unimpaired lungs, you have a healthy animal, ready for the heavier burdens of study. Preserve beyond all else as the priceless portion of a child the integrity of the nervous system. Upon this depends their success in life. With the nervous system shattered, what is life worth? Suppose you begin the education, so-called, of your child at, say, three or four, if he be unusually bright, in the kindergarten. Keep adding slowly and systematically, with what I think the devil must enjoy as a refined means of torment, to the burden day by day. Keep on "educating" him until he enters the primary school at five, and push him to the uttermost until he is ten. You have now laid broad and deep the foundation; outraged nature may be left to take care of the rest.

The integrity of your child's nervous system, no matter what any so-called educator may say, is thus impaired; he can never again be what he would have been had you taken him as the plant-cultivator takes a plant, and for these first ten precious years of his life had fitted him for the future. Nothing else is doing so much to break down the nervous systems of Americans, not even the insane rushing of maturer years, as this over-crowding and cramming of child-life before the age of ten. And the mad haste of maturer years is the legitimate result of the earlier strain.

NEITHER PLANT NOR CHILD TO BE OVERFED

Nor should the child, any more than the plant, be overfed, but more especially should not be given an unbalanced ration. What happens when we overfeed a plant, especially an unbalanced ration? Its root system, its leaf system, its trunk, its whole body, is impaired. It becomes engorged. Following this, comes devitalization. It is open to attacks of disease. It will easily be assailed by fungous diseases and insect pests. It rapidly and abnormally grows onward to its death. So with a child you can easily over-feed it on an unbalanced ration, and the result will be as disastrous as in the case of the plant. The effect of such an unbalanced ration as that fed to the children in the community I have referred to was to shorten life; they developed prematurely, and died early.

Again some one says, But how can the very poor feed their children plenty of nutritious food?

I answer that the nation must protect itself. I mean by this that it is imperative, in order that the nation may rise to its full powers and accomplish its destiny, that the people who comprise this nation must be normal physically. It is imperative, in order that the nation be normal, that the plants of the nation from which it derives its life and without which the nation dies must be sound. All human life is absolutely dependent upon plant life. If the plant life be in any measure lowered through lack of nourishment, with the inevitable lack of ability to produce the best results, the nation suffers. To the extent that any portion of the people are physically mentally or morally unfit, to that extent the nation is weakened.

Do not misunderstand me: I am not advocating paternalism in any sense; far from it. But is not the human race worth as much care as the orchards, the farms, the cattle-ranges? I would so work upon this great blending of races, upon each individual factor in it, that each factor should be called upon to do its very best, be compelled to do its very best, if it was shirking responsibility. But in any great nation there must be a large number who cannot do their best, if I may use a contradictory term, who do not seem able to rise to their opportunities and their possibilities. Already you may see in our larger cities efforts in a small way to help feed the very poor. It can be done nationally as well as municipally, and it can be done so that no loss of self-respect will follow, no encouragement and fostering of poverty or laziness.

Then, too, there are the orphans and the waifs; these must be taken into account. They must have wise, sane, consistent state aid. I am opposed to all sectarian aid. I would do away with all asylums of all types for the indigent under sectarian or private control. The nation, or the commonwealth, should take care of the unfortunate. It must do this in a broad and liberal and sane manner, if we are ever to accomplish the end sought, to make this nation rise to its possibilities. Only through the nation, or State, can this work be done. It must be done for self-protection.

V

DANGERS

In the immediate future, possibly within your life and mine, unquestionably within the life of this generation, what have we most to fear in America from this vast crossing of races? Not in the vicious adults who are now with us, for they can be controlled by law and force, but in the children of these adults, when they have grown and been trained to responsible age in vice and crime, lies the danger. We must begin now, to-day, the work of training these children as they come. Grant that it were possible that every boy and girl born in the United States during the next thirty years should be kept in an atmosphere of crime to the age of ten. The result would be too appalling to contemplate. As they came to adult years, vice would be rampant, crime would go unpunished, all evil would thrive, the nation would be destroyed. Now, to the extent that we leave the children of the poor and these other unfortunates,—waifs and foundlings,—to themselves and their evil surroundings, to that extent we breed peril for ourselves.

The only way to obviate this is absolutely to cut loose from all precedent and begin systematic State and National aid, not next year, or a decade from now, but to-day. Begin training these outcasts, begin the cultivation of them, if you will, much as we cultivate the plants, in order that their lives may be turned into right ways, in order that the integrity of the state may be maintained. Rightly cultivated, these children may be made a blessing to the race; trained in the wrong way, or neglected entirely, they will become a curse to the state.

ENVIRONMENT

LET us bring the application still nearer home.

There is not a single desirable attribute which, lacking in a plant, may not be bred into it. Choose what improvement you wish in a flower, a fruit, or a tree, and by crossing, selection, cultivation, and persistence you can fix this desirable trait irrevocably. Pick out any trait you want in your child, granted that he is a normal child,—I shall speak of the abnormal later,—be it honesty, fairness, purity, lovableness, industry, thrift, what not. By surrounding this child with sunshine from the sky and your own heart, by giving the closest communion with nature, by feeding this child well-balanced, nutritious food, by giving it all that is implied in healthful environmental influences, and by doing all in love, you can thus cultivate in the child and fix there for all its life all of these traits. Naturally not always to the full in all cases at the beginning of the work, for heredity will make itself felt first, and, as in the plant under improvement, there will be certain strong tendencies to reversion to former ancestral traits; but, in the main, with the normal child, you can give him all these traits by patiently, persistently, guiding him in these early formative years.

And, on the other side, give him foul air to breathe, keep him in a dusty factory or an unwholesome school-room or a crowded tenement up under the hot roof; keep him away from the sunshine, take away from him music and laughter and happy faces; cram his little brains with so-called knowledge, all the more deceptive and dangerous because made so apparently adaptable to his young mind; let him have vicious associates in his hours out of school, and at the age of ten you have fixed in him the opposite traits. He is on his way to the gallows. You have perhaps seen a prairie fire sweep through the tall grass across a plain. Nothing can stand before it, it must burn itself out. That is what happens when you let the weeds grow up in a child's life, and then set fire to them by wrong environment.

THE ABNORMAL

BUT, some one asks, What will you do with those who are abnormal? First, I must repeat that the end will not be reached at a bound. It will take years, centuries, perhaps, to erect on this great foundation we now have in America the structure which I believe is to be built. So we must begin to-day in our own commonwealth, in our own city or town, in our own family, with ourselves. Here appears a child plainly not normal, what shall we do with him? Shall we, as some have advocated, even from Spartan days, hold that the weaklings should be destroyed? No. In cultivating plant life, while we destroy much that is unfit, we are constantly on the lookout for what has been called the abnormal, that which springs apart in new lines. How many plants are there in the world to-day that were not in one sense once abnormalities? No; it is the influence of cultivation, of selection, of surroundings, of environment, that makes the change from the abnormal to the normal. From the children we are led to call abnormal may come, under wise cultivation and training, splendid normal natures. A great force is sometimes needed to change the aspect of minerals and metals. Powerful acids, great heat, electricity, mechanical force, or some such influence, must be brought to bear upon them. Less potent influences will work a complete change in plant-life. Mild heat, sunshine, the atmosphere, and greatly diluted chemicals, will all directly affect the growth of the plant and the production of fruits and flowers. And when we come to animal life, especially in man, we find that the force or influence necessary to affect a transformation is extremely slight. This is why environment plays such an important part in the development of man.

In child-rearing, environment is equally essential with heredity. Mind you, I do not say that heredity is of no consequence. It is the great factor, and often makes environment almost powerless. When certain hereditary tendencies are almost indelibly ingrained, environment will have a hard battle to effect a change in the child; but that a change can be wrought by the surroundings we all know. The particular subject may at first be stubborn against these influences, but repeated application of the same modifying forces in succeeding generations will at last accomplish the desired object in the child as it does in the plant.

No one shall say what great results for the good of the race may not be attained in the cultivation of abnormal children, transforming them into normal ones.

THE PHYSICALLY WEAK

So also of the physically weak. I have a plant in which I see wonderful possibilities, but it is weak. Simply because it is weak do I become discouraged and say it can never be made strong, that it would better be destroyed? Not at all; it may possess other qualities of superlative value. Even if it never becomes as robust as its fellows, it may have a tremendous influence. Because a child is a weakling, should it be put out of the way? Such a principle is monstrous. Look over the long line of the great men of the world, those who have changed history and made history, those who have helped the race upward,—poets, painters, statesmen, scientists, leaders of thought in every department,—and you will find that many of them have been physically weak. No, the theory of the ancients that the good of the state demanded the elimination of the physically weak was, perhaps, unwise. What we should do is to strengthen the weak, cultivate them as we cultivate plants, build them up, make them the very best they are capable of becoming.

THE MENTALLY DEFECTIVE

But with those who are mentally defective—ah, here is the hardest question of all!—what shall be done with them? Apparently fatally deficient, can they ever be other than a burden? In the case of plants in which all tendencies are absolutely vicious there is only one course—they must be destroyed. In the case of human beings in whom the light of reason does not burn, those who, apparently, can never be other than a burden, shall they be eliminated from the race? Go to the mother of an imbecile child and get your answer. No; here the analogy must cease. I shall not say that in the ideal state general citizenship would not gain by the absence of such classes, but where is the man who would deal with such Spartan rigor with the race? Besides all this, in the light of the great progress now being made in medical and surgical skill, who shall say what now apparently impossible cures may not be effected?

But it is as clear as sunlight that here,

as in the case of plants, constant cultivation and selection will do away with all this, so that in the grander race of the future these defectives will have become permanently eliminated from the race heredity. For these helpless unfortunates, as with those who are merely unfortunate from environment, I should enlist the best and broadest state aid.

VI

MARRIAGE OF THE PHYSICALLY UNFIT

IT would, if possible, be best absolutely to prohibit in every State in the Union the marriage of the physically, mentally and morally unfit. If we take a plant which we recognize as poisonous and cross it with another which is not poisonous and thus make the wholesome plant evil, so that it menaces all who come in contact with it, this is criminal enough. But suppose we blend together two poisonous plants and make a third even more virulent, a vegetable degenerate, and set their evil descendants adrift to multiply over the earth, are we not distinct foes to the race? What, then, shall we say of two people of absolutely defined physical impairment who are allowed to marry and rear children? It is a crime against the state and every individual in the state. And if these physically degenerate are also morally degenerate, the crime becomes all the more appalling.

COUSINS

WHILE it seems clear now in the light of recent studies that the children of first cousins who have been reared under different environmental influences and who have remained separate from birth until married are not likely to be impaired either mentally, morally or physically, though the second generation will be more than likely to show retrogression, yet first cousin marriages when they have been reared under similar environment should, no doubt, be prohibited. The history of some of the royal families of Europe, where intermarrying, with its fatal results, has so long prevailed, should be sufficient though in these cases other baneful influences have no doubt added their shadow to the picture.

TEN GENERATIONS

But let us take a still closer view of the subject. Suppose it were possible to select say, a dozen normal families, the result of some one of the many blendings of these native and foreign stocks, and let them live by themselves, so far as the application of the principles I have been speaking of are concerned, though not by any means removed from the general influences of the state. Let them have, if you will, ideal conditions for working out these principles, and let them be solemnly bound to the development of these principles—what can be done?

In plant cultivation, under normal conditions, from six to ten generations are generally sufficient to fix the descendants of the parent plants in their new ways. Sufficient time in all cases must elapse so that the descendants will not revert to some former condition of inefficiency. When once stability is secured, usually, as indicated, in from six to ten generations, the plant may then be counted upon to go forward in its new life as though the old lives of its ancestors had never been. This, among plants, will be by the end of from five to ten generations, varying according to the plant's character—its pliability or stubbornness. I do not say that lack of care and nourishment thereafter will not have a demoralizing influence, for no power can prevent a plant from becoming again part wild if left to itself through many generations, but even here it will probably become wild along the lines of its new life, not by any means necessarily along ancestral lines.

If, then, we could have these twelve families under ideal conditions where these principles could be carried out unswervingly, we could accomplish more for the race in ten generations than can now be accomplished in a hundred thousand years. Ten generations of human life should be ample to fix any desired attribute. This is absolutely clear. There is neither theory nor speculation. Given the fact that the most sensitive material in all the world upon which to work is the nature of a little child, given ideal conditions under which to work upon this nature, and the end desired will as certainly come as it comes in the cultivation of the plant. There will be this difference, however, that it will be immeasurably easier to produce and fix any desired traits in the child than in the plant, though, of course, a plant may be said to be a harp with a few strings as compared with a child.

THE PERSONAL ELEMENT

BUT some one says, You fail to take into account the personal element, the sovereign will of the human being, its power of determining for itself.

By no means; I give full weight to this. But the most stubborn and wilful nature in the world is not that of a child. I have dealt with millions of plants, have worked with them for many years, have studied them with the deepest interest from all sides of their lives. The most stubborn living thing in this world, the most difficult to swerve, is a plant once fixed in certain habits—habits which have been intensified and have been growing stronger and stronger upon it by repetition through thousands and thousands of years. Remember that this plant has preserved its individuality all through the ages; perhaps it is one which can be traced backward through eons of time in the very rocks themselves, never having varied to any great extent in all these vast periods. Do you suppose, after all these ages of repetition, the plant does not become possessed of a will, if you so choose to call it, of unparalleled tenacity? Indeed, there are plants, like certain of the palms, so persistent that no human power has yet been able to change them. The human will is a weak thing beside the will of a plant. But see how this whole plant's lifelong stubbornness is broken simply by blending a new life with it, making, by crossing, a complete and powerful change in its life. Then when the break comes, fix it by these generations of patient supervision and selection, and the new plant sets out upon its new way never again to return to the old, its tenacious will broken and changed at last.

When it comes to so sensitive and pliable a thing as the nature of a child, the problem becomes vastly easier.

VII

HEREDITY—PREDESTINATION—TRAINING

There is no such thing in the world, there never has been such a thing, as a predestined child—that is predestined for heaven or hell. Men have taught such things in the past, there may be now those who account for certain manifestations on this belief, just as there may be those who in the presence of some hopelessly vicious man hold to the view, whether they express it or not, of total depravity. But even total depravity never existed in a human being, never can exist in one any more than it can exist in a plant. Heredity means much, but what is heredity? Not some hideous ancestral specter forever crossing the path of a human being. Heredity is simply the sum of all the effects of all the environments of all past generations on the responsive, ever-moving life forces. There is no doubt that if a child with a vicious temper be placed in an environment of peace and quiet the temper will change. Put a boy born of gentle white parents among Indians and he will grow up like an Indian. Let the child born of criminal parents have a setting of morality, integrity, and love, and the chances are that he will not grow into a criminal, but into an upright man. I do not say, of course, that heredity will not sometimes assert itself. When the criminal instinct crops out in a person, it might appear as if environment were leveled to the ground; but in succeeding generations the effect of constant higher environment will not fail to become fixed.

Apply to the descendants of these twelve families throughout three hundred years the principles I have set forth, and the reformation and regeneration of the world, their particular world, will have been effected. Apply these principles now, to-day, not waiting for the end of these three hundred years, not waiting, indeed, for any millennium to come, but *make* the millennium, and see what splendid results will follow. Not the ample results of the larger period, to be sure, for with the human life, as with the plant life, it requires these several generations to fix new characteristics or to intensify old ones. But narrow it still more, apply these principles to a single family,—indeed, still closer, to a single child, your child it may be,—and see what the results will be.

But remember that just as there must be in plant cultivation great patience, unswerving devotion to the truth, the highest motive, absolute honesty, unchanging love, so must it be in the cultivation of a child. If it be worth while to spend ten years upon the ennoblement of a plant, be it fruit, tree, or flower, is it not worth while to spend ten years upon a child in this precious formative period, fitting it for the place it is to occupy in the world? Is not a child's life vastly more precious than the life of a plant? Under the old order of things plants kept on in their course largely uninfluenced in any new direction. The plant-breeder changes their lives to make them better than they ever were before. Here in America, in the midst of this vast crossing of species, we have an unparalleled opportunity to work upon these sensitive human natures. We may surround them with right influences. We may steady them in right ways of living. We may bring to bear upon them, just as we do upon plants, the influence of light and air, of sunshine and abundant, well-balanced food. We may give them music and laughter. We may teach them as we teach the plants to be sturdy and self-reliant. We may be honest with them, as we are obliged to be honest with plants. We may break up this cruel educational articulation which connects the child in the kindergarten with the graduate of the university while there goes on from year to year an uninterrupted system of cramming, an uninterrupted mental strain upon the child, until the integrity of its nervous system may be destroyed and its life impaired.

I may only refer to that mysterious prenatal period, and say that even here we should begin our work, throwing around the mothers of the race every possible loving, helpful, and ennobling influence; for in the doubly sacred time before the birth of a child lies, far more than we can possibly know, the hope of the future of this ideal race which is coming upon this earth if we and our descendants will it so to be.

Man has by no means reached the ultimate. The fittest has not yet arrived. In the process of elimination the weaker must fail, but the battle has changed its base from brute force to mental integrity. We now have what are popularly known as five senses, but there are men of strong minds whose reasoning has rarely been at fault and who are coldly scientific in their methods, who attest to the possibility of yet developing a sixth sense. Who is he who can say man will not develop new senses as evolution advances? Psychology is now studied in most of the higher institutions of learning throughout the country, and that study will lead to a greater knowledge of these subjects. The man of the future ages will prove a somewhat different order of being from that of the present. He may look upon us as we today look upon our ancestors.

Statistics show many things to make us pause, but, after all, the only right and proper point of view is that of the optimist. The time will come when insanity will be reduced, suicides and murders will be greatly diminished, and man will become a being of fewer mental troubles and bodily ills. Whenever you have a nation in which there is no variation, there is comparatively little insanity or crime, or exalted morality or genius. Here in America, where the variation is greatest, statistics show a greater percentage of all these variations.

As time goes on in its endless and ceaseless course, environment must crystallize the American nation; its varying elements will become unified, and the weeding-out process will, by the means indicated in this paper, by selection and environmental influences, leave the finest human product ever known. The transcendent qualities which are placed in plants will have their analogies in the noble composite, the American of the future.

VIII

GROWTH

GROWTH is a vital process—an evolution—a marshaling of vagrant unorganized forces into definite forms of beauty, harmony and utility. Growth in some form is about all that we ever take any interest in; it expresses about everything of value to us. Growth in its more simple or most marvelously complicated forms is the architect of beauty, the inspiration of poetry, the builder and sustainer of life, for life itself is only growth, an ever-changing movement toward some object or ideal. Wherever life is found, there, also, is growth in some direction. The end of growth is the beginning of decay.

Growth within, is health, content and happiness, and growing things without stimulate and enhance growth within. Whose pulses are not hastened, and who is not filled with joy when in Earth's long circling swing around our great dynamo the Sun, the point is reached where chilling, blistering frosts are exchanged for warmth and growth! When the flowers and grasses on the warm hillsides gleefully hasten up through the soft wet soil, or later when ferns, meadow rues and trilliums thrilled with awakened life, crack through and push up the loose mellow earth in small mounds —little volcanoes of growth; all these variously organized life forces are expressing themselves each in its own specific way. Each so-called species, each individual has something within itself which we call heredity—a general tendency to reproduce itself in form and habits somewhat definitely after its own kind.

NEW SPECIES

Most of the ancient and even a large part of modern students of plant and animal life have held that their so-called true species never varied to any great extent, at least never varied from the standard type sufficiently to form what could scientifically be called a new species. Under this view the word heredity has had a very indefinite meaning when used in conjunction with environment; and a never-ending uncertainty has always been apparent as to their relative power in molding individual life.

HEREDITY AND ENVIRONMENT

WHEN the great rivers of life, which we now see, commenced on this planet they did not at once leap into existence with all their present complicated combinations of forces and motions; all were very insignificant; their slender courses, though simple, were devious and uncertain, at first lacking all the wonderfully varied but slowly acquired adaptations to environment that have come with the ages; all had many obstacles to overcome, many things to learn;—and for long ages were able to respond only to the more powerful or long-continued action of external forces. Many of these frail life streams in the long race down the ages were snuffed out by unfavorable surroundings, unfavorable heredity, or the combination and interaction of both; others more successful have lived to be our contemporaries and to-day the process is still unchanged.

If a race has not acquired and stored among its hereditary tendencies sufficient perseverance and adaptability to meet all the changes to which it must always be subjected by its ever-changing environment, it will be left behind and finally destroyed, outstripped by races better equipped for the fray.

IX

ENVIRONMENT THE ARCHITECT OF HEREDITY

HEREDITY is not the dark specter which some people have thought —merciless and unchangeable, the embodiment of Fate itself. This dark, pessimistic belief which tinges even the literature of to-day comes, no doubt, from the general lack of knowledge of the laws governing the interaction of these two ever-present forces of heredity and environment wherever there is life.

My own studies have led me to be assured that heredity is only the sum of all past environment, in other words environment is the architect of heredity; and I am assured of another fact: acquired characters *are* transmitted and—even further—that *all* characters which *are* transmitted have been acquired, not necessarily at once in a dynamic or visible form, but as an increasing latent force ready to appear as a tangible character when by long-continued natural or artificial repetition any specific tendency has become inherent, inbred, or "fixed," as we call it.

We may compare this sum of the life forces, which we call heredity, to the character of a sensitive plate in the camera. Outside pictures impress themselves more or less distinctly on the sensitive plate according to their position, intensity, and the number of times the plate has been exposed to the objects (environments) in the same relative position; all impressions are recorded. Old ones fade from immediate consciousness, but each has written a permanent record. Stored within heredity are all joys, sorrows, loves, hates, music, art, temples, palaces, pyramids, hovels, kings, queens, paupers, bards, prophets and philosophers, oceans, caves, volcanoes, floods, earthquakes, wars, triumphs, defeats, reverence, courage, wisdom, virtue, love and beauty, time, space, and all the mysteries of the universe. The appropriate environments will bring out and intensify all these general human hereditary experiences and quicken them again into life and action, thus modifying for good or evil character—heredity—destiny.

REPETITION

Repetition is the best means of impressing any one point on the human understanding; it is also the means which we employ to train animals to do as we wish, and by just the same process we impress plant life. By repetition we fix any tendency, and the more times any unusual environment is repeated the more indelibly will the resultant tendencies be fixed in plant, animal, or man, until, if repeated often enough in any certain direction, the habits become so fixed and inherent in heredity that it will require many repetitions of an opposite nature to efface them.

APPLICATION TO CHILD LIFE

WHAT possibilities this view opens up in the culture and development of the most sensitive and most precious of all lives which ever come under our care and culture—child life!

Can we hope for normal, healthy, happy children if they are constantly in ugly environment? Are we not reasonably sure that these conditions will almost swamp a well-balanced normal heredity and utterly overthrow and destroy a weak though otherwise good one?

We are learning that child life is far more sensitive to impressions of any kind than we had ever before realized, and it is certain that this wonderful sensitiveness and ready adaptability has not as yet by any means been put to its best possible use in child culture—either in the home or the school—and though all must admire our great educational system, yet no well-informed person need be told that it is not perfect.

X

CHARACTER

WE are a garrulous people and too often forget, or do not know, that the heart as well as the head should receive its full share of culture. Much of our education has been that of the parrot; children's minds are too often crowded with rules and words. Education of the intellect has its place, but is injurious, unnatural, and unbalanced unless in addition to cultivating the memory and reason we educate the heart also in the truest sense. A well-balanced character should always be the object and aim of all education.

A perfect system of education can never be attained because education is preparing one for the environment expected, and conditions change with time and place. There is too much striving to be consistent rather than trying to be right. We must learn that what we call character is heredity and environment in combination, and heredity being only *stored environment* our duty and our privilege is to make the stored environment of the best quality; in this way character is not only improved in the individual but the desired qualities are added to heredity to have their influence in guiding the slightly but surely changed heredities of succeeding generations.

SUCCESS

COLD mathematical intellect unaccompanied by a heart for the philosophic, idealistic, and poetic side of nature is like a locomotive well made but of no practical value without fire and steam; a good knowledge of language, history, geography, mathematics, chemistry, botany, astronomy, geology, etc., is of some importance, but far more so is the knowledge that all true success in life depends on integrity; that health, peace, happiness, and content, all come with heartily accepting and daily living by the "Golden Rule"; that dollars, though of great importance and value, do not necessarily make one wealthy; that a loving devotion to truth is a normal indication of physical and mental health; that hypocrisy and deceit are only forms of debility, mental imbecility and bodily disease, and that the knowledge and ability to perform useful, honest labor of any kind is of infinitely more importance and value than all the so-called "culture" of the schools, which too often turn out nervous pedantic victims of unbalanced education with plenty of words but with no intuitive ability to grasp, digest, assimilate and make use of the environment which they are compelled each day to meet and to conquer or be conquered.

Any form of education which leaves one less able to meet every-day emergencies and occurrences is unbalanced and vicious, and will lead any people to destruction.

Every child should have mud pies, grasshoppers, water-bugs, tadpoles, frogs, mud-turtles, elderberries, wild strawberries, acorns, chestnuts, trees to climb, brooks to wade in, water-lilies, woodchucks, bats, bees, butterflies, various animals to pet, hay-fields, pine-cones, rocks to roll, sand, snakes, huckleberries and hornets; and any child who has been deprived of these has been deprived of the best part of his education.

By being well acquainted with all these they come into most intimate harmony with nature, whose lessons are, of course, natural and wholesome.

A fragrant beehive or a plump, healthy hornet's nest in good running order often become object lessons of some importance. The inhabitants can give the child pointed lessons in punctuation as well as caution and some of the limitations as well as the grand possibilities of life; and by even a brief experience with a good patch of healthy nettles, the same lesson will be still further impressed upon them. And thus by each new experience with homely natural objects the child learns self-respect and also to respect the objects and forces which must be met.

XI

FUNDAMENTAL PRINCIPLES

"KNOWLEDGE is Power," but it requires to be combined with wisdom to become useful. The fundamental principles of education should be the subject of earnest scientific investigation, but this investigation should be broad, including not only the theatrical, wordy, memorizing, compiling methods, but should also include *all* the causes which tend to produce men and women with sane well-balanced characters.

We must learn that any person who will not accept what he knows to be truth, for the very love of truth alone, is very definitely undermining his mental integrity. It will be observed that the mind of such a person gradually stops growing, for, being constantly hedged in and cropped here and there, it soon learns to respect artificial fences more than freedom for growth. You have not been a very close observer of such men if you have not seen them shrivel, become commonplace, mean, without influence, without friends and the enthusiasm of youth and growth, like a tree covered with fungus, the foliage diseased, and the life gone out of the heart with dry rot and indelibly marked for destruction—dead, but not yet handed over to the undertaker.

The man or the woman who moves the earth, who is master rather than the victim of fate, has strong feelings well in hand—a vigilant engineer at the throttle.

Education which makes us lazier and more helpless is of no use. Leaders use the power within; it should give the best organized thought and experience of men through all the ages of the past. By it we should learn that it is not necessary to be selfish in order to succeed. If you happen to get a new idea don't build a barbed wire fence around it and label it yours. By giving your best thoughts freely others will come to you so freely that you will soon never think of fencing them in. Thoughts refuse to climb barbed wire fences to reach anybody.

By placing ourselves in harmony and coöperation with the main high potential line of human progress and welfare we receive the benefit of strong magnetic induction currents. But by placing our life energies at right angles to it we soon find ourselves on a low-feed induction current, thus losing the help and support which should be ours.

Straightforward honesty always pays better dividends than zigzag policy. It gives one individuality, self-respect, and power to take the initiative, saving all the trouble of constant tacking to catch the popular breeze. Each human being is like a steamship, endowed with a tremendous power. The fires of life develop a pressure of steam which, well disciplined, leads to happiness for ourselves and others; or it may lead only to pain and destruction.

To guide these energies is the work

of true education. Education of rules and words only for polish and public opinion is of the past. The education of the present and future is to guide these energies through wind and wave straight to the port desired. Education gives no one any new force. It can only discipline nature's energies to develop in natural and useful directions so that the voyage of life may be a useful and happy one—so that life may not be blasted or completely cut off before thought and experience have ripened into useful fruit.

When the love of truth for truth's sake—this poetic idealism, this intuitive perception, this growth from within—has been awakened and cultivated, thoughts live and are transmitted into endless forms of beauty and utility;

but to receive this new growth we must cultivate a sturdy self-respect, we must break away from the mere petrified word-pictures of others and cultivate the "still small voice" within by which we become strong in individual thought and quick in action, not cropped, hedged and distorted by outward, trivial forms, fads and fancies. Every great man or woman is at heart a poet, and all must listen long to the harmonies of Nature before they can make translations from her infinite resources through their own ideals into creations of beauty in words, forms, colors, or sounds. Mathematical details are invaluable, the compilation method is beyond reproach; intellectually we may know many things, but they will never be of any great value toward a normal growth unless there is an inward awakening, an intuitive grasp, an impelling personal force which digests, assimilates and individualizes. This intuitive consciousness, combined with extensive practical knowledge and "horse sense," has always been the motive power of all those who have for all time left the human race rich with legacies of useful thought, with ripening harvests of freedom and with ever-increasing stores of wisdom and happiness. We are now standing upon the threshold of new methods and new discoveries which shall give us imperial dominion.